S0-DOO-198
Crafty Cans
From The Kitchen
to:
from:
Cake & Bread Mixes To Give As Gifts In Cans!
Decorate The Cans With Paper Or Fabric, Rubber Stamps,
Stickers, Bows Or Lace - However You Like!
The Recipient Bakes It In The Same Can!
Jackie Gannaway

Published in Austin, TX by COOKBOOK CUPBOARD
P.O. Box 50053, Austin, TX 78763 (512) 477-7070 phone (512) 891-0094 fax

ISBN 1-885597-40-1

Artwork by Frank Bielec of Mosey 'N Me - Katy, TX.

Over 1.5 Million Copies Of Jackie Gannaway's Books Are In Print!

This book is part of
"Jackie's Originals Collection-
Creative New Concepts in Gift Mixes"™

Thanks to PrintMaster software, published by Broderbund, for copyright permission to reproduce some of their copyright images (photos - pg. 7)

Thanks to Ann Trask, Austin, TX, for her help with computer clip art for can decorations. (See Ann in the top left hand photo - pg. 7)

Mail Order Information

To order a copy of this book send a check for $3.95 + $1.50 for shipping (TX residents add 8.25 % sales tax) to Cookbook Cupboard, P.O. Box 50053, Austin, TX 78763. Send a note asking for this title by name. If you would like a descriptive list of the nearly 40 fun titles in The Kitchen Crafts Collection an The Layers of Love Collection™ send a note, call or fax (numbers at top of this page) asking for a brochure.

What Is This Book About?

This book has recipes for a brand new way to give a cake or bread mix as a gift or sell at a bazaar - (see note on copyright page). Save all your food cans (14 to 16 oz.) and recycle them in this fun way instead of putting the cans out at the curb on recycling day.

Follow these recipes and put a packet of cake or bread mix and a packet of glaze or frosting mix into a can.

Decorate the can by replacing the existing label from the can with your own art or crafts. The decorations can be as simple (brown paper sack, wrapping paper) or as elaborate as you want - many, many ideas are listed on pgs. 4 - 5.

This is perfect for scrapbookers because all the leftover scrapbooking supplies can be put to good use this way. It is also perfect for rubber stampers because the base is a paper label that can be stamped, embossed, personalized many ways with rubber stamped designs.

Those of you who work with fabric, lace and beads can use your small leftover pieces to decorate the cans.

People who keep family photographs in their computers can print out those photographs with a message (Happy Birthday, Grandma, etc.) and make labels for the cans that way.

The recipient adds an egg, water and oil to the mix - puts the batter right back in the same can you gave them - puts the can in the oven - and bakes the cake or bread.

The small baked cake or bread will be the size of the can. It can be sliced into 4 to 8 round slices, depending on thickness.

While the cake is baking, the recipient will add water or butter to the glaze or frosting mix and will be all ready to sweeten the slices with some of the glaze or frosting.

This is very easy for you as well as for the recipient. It is very inexpensive to make. The ingredients needed by the recipient are common ingredients in every kitchen.

I hope you have fun with my brand new idea for giving mixes in a new, different way. Read all the basic information and tips below and on pgs. 4 - 6. See the photos on page. 7.

A complete list of recipes is in the Index on page 32. *Jackie Gannaway*

Recipe Cards

Write instructions onto a recipe card so the recipient will know how to turn the mix into a baked cake or bread. The words to use are written inside a box at the bottom of each recipe in this book.

Coordinate the recipe card with the decorations on the can. Tuck the recipe card inside the can with the mix. (You can attach it to the outside of the can if it won't take away from your can decorations.)

Use scalloped or fancy-edge scissors to cut the recipe card from heavy paper. Use the same scissors to cut out the "To-From" cards.

All About Cans

For the recipes in this book you need cans from food (like corn, green beans, etc.). Food comes in many size cans. The can size you need is between 14 oz. to 16 oz. It will say that on the label. A larger or smaller can will not work for these recipes. This is a very common size can and very easy to find.

Before using the can, take pliers (small needle - nose pliers work best) and tightly squeeze the place inside the can lip where the can opener first cut into the can. This keeps any metal from scratching you or the recipient.

If you don't personally use enough food cans, ask your friends and family to save their cans for you. Ask your neighbors who regularly recycle to save their cans for you. This is even better than recycling - it is RE-USING. The can will serve two purposes in its time, and then could still be recycled by the recipient after she has baked the cake or bread in it.

If you look for a source from which to buy large quantities of cans be sure the cans are food safe. You know they are food safe if you re-use the cans corn and green beans originally came in.

These mixes are non-breakable and fairly lightweight. This makes them perfect for mailing to out of town friends and family.

General Can Decorating Information

Each 8 1/2" x 11" sheet of scrapbooking paper, computer paper with a pretty design, plain colored paper, construction paper, etc. will make two labels for cans.

Get your pattern for the label by removing the label from the can and measuring it. It will be approximately 4 1/8" x 10". If you are using 8 1/2" x 11" paper, cut it in half lengthwise for two labels. 11" is a little longer than you need - just overlap it further, or trim it. The easiest way to hold the label to the can is with double stick tape. Just wrap the label around the can - overlap it and place the double stick tape at that place. You can also glue the ends of the label together, but double stick tape is by far the easiest way.

DO NOT GLUE THE LABEL TO THE ENTIRE CAN SURFACE - don't decorate the can with contact paper. The reason for this is that the person is going to bake right in the can in their oven. They will have to remove the label and any attached decorations before they bake. If you have personalized the label they can keep it as a souvenir.

Once you have your label - add anything you want to the label. Use such things as stickers, fabric fringe, glued on "jewels", photos. Or the label itself can be drawn by hand by a child or adult, designed with a clip art program on a computer and personalized to the recipient ("Merry Christmas to Grandma from Susie and Donnie"). It can also be personalized for the fund-raiser ("Can Cakes From Washington Elementary" - and decorated in the school colors).

Stick-On Labels And "To-From" Cards

The glaze and frosting mixes need to be labeled. Use plain white office "peel and stick" labels and hand write on them. Or print out labels with clip art from your computer. Or buy colored office labels to coordinate with the decorations on your can.

You also need to attach a "To - From" card. Another opportunity to be creative. There are as many opportunities as anyone might want to coordinate the can with the label with the cellophane with the instruction cards with the stick-on labels with the To-From cards. Have fun! Be as creative or as simple as you want.

"Finishing" The Can Decorations

The can will look more professionally "finished" if you place the bags of mix inside colored tissue paper or colored cellophane and then push them inside the can. Allow about 4" to 6" of tissue paper or cellophane to stick out of the top of the can (like it sticks out of a gift bag for a splash or color). OR use shredded Easter "grass" in a coordinating color. Place it in the can on top of the bags of mix. This gives a festive look. (See photos - pg. 7.)

"Stand-Up" Can Decorations

Use heavy paper and your computer to print out a 6" to 9" tall picture (a birthday cake with candles, a Christmas tree, a photo of your child). Cut out the shape and discard the remaining paper. If you don't work with computers you can find heavy paper cut-out designs for sale at craft stores.

Use this cut-out either behind or in front of can. The can will stand up on the table with a tall design attached to the back or front of the can and sticking up 4" to 6" taller than the can. This is a dramatic look. Attach it to the paper label with hot glue or extra tape. (See photos - pg. 7.)

Specific Can Decorating Information

Use the following ideas as ways to decorate cans: These ideas are good for any skill level - from toddler to the most accomplished artist or crafter or quilter.

Brown paper sacks.
Stickers.
Bumper stickers.
Stencils.
Postcards.
Gift wrapping paper.
Bows and ribbons.
Tinsel garlands.

Magazine pages - catalog pages.
Wallpaper borders - leftover wallpaper.
Recycled greeting cards- Christmas cards.
Glue on shapes made with punches.
Write with glitter pens, paint pens, etc.
College print tissue paper.
Jingle bells or other seasonal accessories.
Coloring book pages.
Make a "doll" - see pg. 32.

Rubber stamp an all-over design on brown kraft paper or white butcher paper - use this for the can label AND to stick out the top of can.
Stencil with watered down white glue - add glitter to the glue.
Make label from a piece of fabric - add ribbons and lace.
Magnets made by children (magnets stick to cans).
Happy birthday wrapping paper - include a birthday candle.
Embossing (use embossing tools or templates from the crafts store.)
Cut labels and decorations with scalloped or decorative-edged scissors.
Glue on scrapbooking paper "photo frames" with a photo inside the frame (make copies or print out photos from your digital camera and computer- don't use original photos)
Print photos and add personalized text from your digital camera and computer.
Children draw pictures with crayons or help with any of the other decorating methods listed here.
Use printed paper napkins for the label and for the "tissue" coming out of the top of the can.
Use 4" long "flapper girl fringe" around the can with a solid color paper label underneath.
Make the entire label of extra wide ribbon, solid or printed. You may have to go around twice to cover entire can if the ribbon is not wide enough.

Make Up Your Own Mix Recipes For Can Cakes

Several of the recipes in this book follow patterns I developed for mixes to bake in cans. If you stick to those patterns and just vary the flavors you should have success. When you experiment with these recipes bake one yourself before giving away the mix to be sure it works. Basic recipes below:

First Basic Cake Mix Recipe

Blend a boxed cake mix (Betty Crocker, Duncan Hines, etc.) with a 4-serving size box of pudding mix. Use any flavors that appeal to you. Place 3/4 cup of this mixture into a zipper sandwich bag.

You will get 6 bags of mix from the entire batch. The recipient adds 1 Tb. water, 1 Tb. oil and 1 egg. Bake in a greased can for 28 to 32 minutes at 350°.

Second Basic Cake Mix Recipe

For each mix use 3/4 cup cake mix (Betty Crocker, Pillsbury, etc.) and 2 Tb. each of two different finely chopped add-ins (nuts, raisins, dried fruit bits, gumdrops, sunflower seeds for a nature bread). DO NOT USE chocolate chips or any other flavor of these kind of chips. They will not work in a can cake. The recipient adds 1 Tb. water, 1 Tb. oil and 1 egg. Bake in a greased can for 28 to 32 minutes at 350°. Make a coordinating glaze (see bottom of page).

Basic Bisquick Recipe

For each mix use 2/3 cup Bisquick, 1 Tb. finely chopped nuts or dried fruit bits and 1/2 tsp. spice. For a sweet bread also add 1 to 2 Tb. sugar or brown sugar. Experiment with combinations.

The recipient adds 3 Tb. milk and 1 egg white. Bake in a greased can 28 to 32 minutes at 375°.

Basic Glaze Recipe

For a vanilla glaze just use 3/4 cup powdered sugar.
The recipient adds 1 to 1 1/2 Tb. water.

For a chocolate glaze add 2 tsp. cocoa powder to the powdered sugar.
For a spice glaze (like cinnamon, nutmeg, ginger) add 1/2 tsp. ground spice to powdered sugar.
For a lemon, lime, orange, cherry or strawberry glaze add 1 Tb. dry drink mix with sugar- like Tang, Country Time lemonade powder, any fruit flavored drink mix with sugar.

Examples Of Just A Few Of The Many Ways To Decorate Cans

Clip art images marked with ** are from PrintMaster software, published by Broderbund.

Personalized Computer
Printed Photo

Personalized Coputer
Printed Photo

College Bumper Sticker
College Tissue Paper

Paper Label, Tinsel
Star Garland

"Doll" **Cut-Out
With Lace Skirt
(see pg. 32)

"Flapper Fringe"
Gold Cord
Cellophane

Paper Christmas Tree
Cut-Out **Attached
To Front Of Can

Tall Candy Cane**
Cut-Out Attached
To Front Of Can

Checked Paper
Decorative Label
Grosgrain Ribbon

All-Over Rubber
Stamp Design On
Brown Kraft Paper

Fabric, Wired Ribbon,
Dimensional
Paper Flower

Patterned Paper
Napkin -Satin
Ribbon

Computer ClipArt**
"Summer Vacation To
Teacher" Message
Attached Paper Star

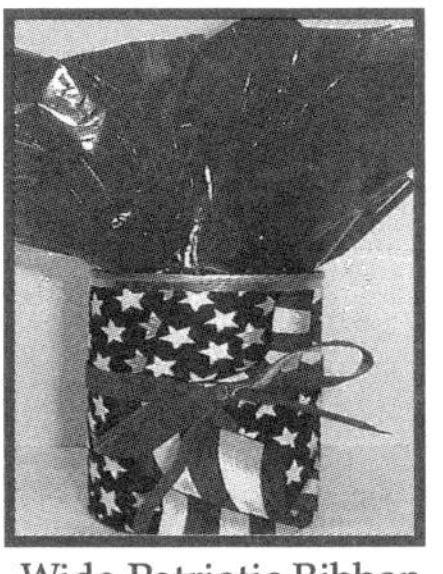

Wide Patriotic Ribbon,
Small
Grosgrain Bow,
Cellophane

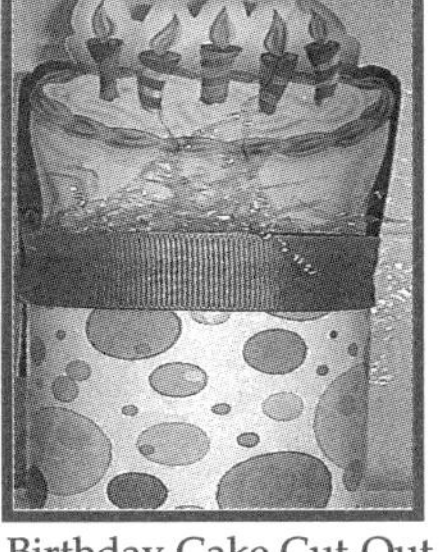

Birthday Cake Cut-Out
On Back Of Can
Wrapping Paper
Easter Grass

Bandana Fabric
Chenille Stem
Jingle Bell
Excelsior

Lemon Cake Mix In A Can

1 lemon cake mix (Betty Crocker, Duncan Hines, etc.)

1 (4-serving size) lemon instant pudding mix

1. Place ingredients into large bowl. Blend with whisk.
2. Measure 3/4 cup of this mixture into a zipper sandwich bag. Repeat. (You should get 6 bags of mix from this cake mix/ pudding mix blend.)

Lemon Glaze Mix

(Make 1 Glaze Mix for each Lemon Cake Mix.)

3/4 cup powdered sugar

1 Tb. lemonade powder with sugar (like Country Time)

3. Place glaze mix ingredients into a small bowl. Blend well. Place in a zipper sandwich bag. Label this bag "Lemon Glaze Mix".
4. Place a bag of cake mix and a bag of glaze mix into each of six 14 to 16 oz. cans (see pgs. 4-5 for detailed instructions on kinds of cans and decorating cans.)
5. Attach a recipe card with the following instructions:

Lemon Cake To Bake In A Can
(With Lemon Glaze)

1. Remove mix from can. (Set aside glaze mix packet for later use.) Empty bag of mix into a medium bowl.
2. Add 1 Tb. water, 1 Tb. oil and 1 egg. Mix well.
3. Remove any labels or decorations from can. Grease can well with some shortening or margarine on a paper towel. (Don't spray can with cooking spray - recipe will not work in a sprayed can.)
4. Spoon batter into can. Bake 28 to 32 minutes at 350°. Accurate oven temp is important. Cake is done when a broom straw or bamboo skewer placed in middle of cake comes out with no wet batter sticking to it.
5. While cake is baking, empty glaze mix into a small bowl. Add 1 to 1 1/2 Tb. water to glaze mix. Stir well.
6. When cake is done let it cool 15 minutes in can. Run plain table knife around edge of cake to loosen it from can. Shake can to allow cake to slide out.
7. To serve, cut cake into slices the desired thickness. Top each serving with some of the glaze.

Orange Slice Cake Mix In A Can

3/4 cup orange cake mix, (Duncan Hines, etc.)

3 orange slice candies (gumdrop type candy called "orange slices")

1. Cut candies into 6 small pieces each. Mix candies with cake mix.
2. Place mixture into a zipper sandwich bag. Place bag into a 14 to 16 oz. can (see pgs. 4-5 for detailed instructions on kinds of cans and decorating cans.)
3. Attach a recipe card with the following instructions:

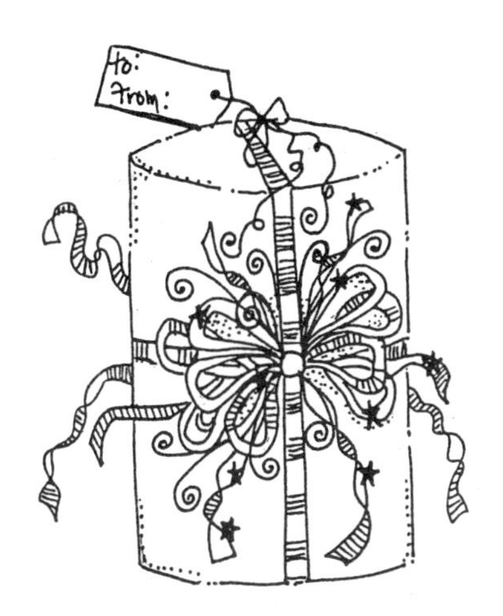

Orange Slice Cake To Bake In A Can

1. Remove mix from can. Empty bag of mix into a medium bowl.
2. Add 1 Tb. water, 1 Tb. oil and 1 egg. Mix well.
3. Remove any labels or decorations from can. Grease can well with some shortening or margarine on a paper towel. (Don't spray can with cooking spray - recipe will not work in a sprayed can.)
4. Spoon batter into can. Bake 32 to 35 minutes at 350°. Accurate oven temp is important. Cake is done when a broom straw or bamboo skewer placed in middle of cake comes out with no wet batter sticking to it.
5. When cake is done let it cool 15 minutes in can. Run plain table knife around edge of cake to loosen it from can. Shake can to allow cake to slide out.
6. To serve, cut cake into slices the desired thickness.

Spice Bread Mix In A Can

2/3 cup Bisquick
2 Tb. brown sugar
2 Tb. finely chopped nuts
1 Tb. sugar
1/2 tsp. cinnamon
1/2 tsp. nutmeg
1/2 tsp. allspice

1. Place ingredients into large bowl. Blend with a whisk.
2. Place mixture into a zipper sandwich bag.
3. Place a bag of bread mix into a 14 to 16 oz. can (see pgs. 4-5 for detailed instructions on kinds of cans and decorating cans.)
5. Attach a recipe card with the following instructions:

Spice Bread To Bake In A Can

1. Remove mix from can. Empty bag of mix into a medium bowl.
2. Add 2 Tb. water, 2 Tb. oil and 1 egg. Mix well.
3. Remove any labels or decorations from can. Grease can well with some shortening or margarine on a paper towel. (Don't spray can with cooking spray - recipe will not work in a sprayed can.)
4. Spoon batter into can. Bake 28 to 32 minutes at 375°. Accurate oven temp is important. Bread is done when a broom straw or bamboo skewer placed in middle of bread comes out with no wet batter sticking to it.
5. When bread is done let it cool 15 minutes in can. Run plain table knife around edge of bread to loosen it from can. Shake can to allow bread to slide out.
6. To serve, cut bread into slices the desired thickness.

Honey Bun Cake Mix In A Can

3/4 cup yellow cake mix (Betty Crocker, Pillsbury, etc.)
1 Tb. brown sugar
1 Tb. finely chopped pecans
1/2 tsp. cinnamon

1. Place cake mix into a zipper sandwich bag.
2. Mix next three ingredients. Place this mixture into a separate zipper bag. Label this "Streusel Mix".

Cinnamon Glaze Mix

(Make 1 Glaze Mix for each Honey Bun Cake Mix.)

3/4 cup powdered sugar
1/2 tsp. cinnamon

3. Place glaze mix ingredients into a small bowl. Blend well. Place in a zipper sandwich bag. Label this bag "Glaze Mix".
4. Place a bag of cake mix, a bag of streusel mix and a bag of glaze mix into each of six 14 to 16 oz. cans (see pgs. 4-5 for detailed instructions on kinds of cans and decorating cans.)
5. Attach a recipe card with the following instructions:

Honey Bun Cake To Bake In A Can (With Cinnamon Glaze)

1. Remove mix from can. (Set aside glaze mix packet for later use.) Empty bag of mix into a medium bowl.
2. Add 1 Tb. water, 1 Tb. oil and 1 egg. Mix well.
3. Sprinkle streusel mix from streusel mix packet into batter. Stir once or twice to just barely mix.
4. Remove any labels or decorations from can. Grease can well with some shortening or margarine on a paper towel. (Don't spray can with cooking spray - recipe will not work in a sprayed can.)
5. Spoon batter into can. Bake 28 to 32 minutes at 350°. Accurate oven temp is important. Cake is done when a broom straw or bamboo skewer placed in middle of cake comes out with no wet batter sticking to it.
6. While cake is baking, empty glaze mix into a small bowl. Add 1 to 1 1/2 Tb. water to glaze mix. Stir well.
7. When cake is done let it cool 15 minutes in can. Run plain table knife around edge of cake to loosen it from can. Shake can to allow cake to slide out.
8. To serve, cut cake into slices the desired thickness. Top each serving with some of the glaze.

Coconut Cake Mix In A Can

1 white cake mix (Betty Crocker, Duncan Hines, etc.)
1 cup flaked coconut
1 (4-serving size) coconut cream instant pudding mix

1. Place ingredients into large bowl. Blend with whisk.
2. Measure 3/4 cup of this mixture into a zipper sandwich bag. Repeat. (You should get 6 bags of mix from this cake mix/pudding mix blend.)

Coconut Frosting Mix

(Make 1 Frosting Mix for each Coconut Cake Mix.)

3/4 cup powdered sugar
1/3 cup flaked coconut

3. Place frosting mix ingredients into a small bowl. Blend well. Place in a zipper sandwich bag. Label this bag "Coconut Frosting Mix".
4. Place a bag of cake mix and a bag of frosting mix into each of six 14 to 16 oz. cans (see pgs. 4-5 for detailed instructions on kinds of cans and decorating cans.)
5. Attach a recipe card with the following instructions:

Coconut Cake To Bake In A Can
(With Coconut Frosting)

1. Remove mix from can. (Set aside frosting mix packet for later use.) Empty bag of mix into a medium bowl.
2. Add 1 Tb. water, 1 Tb. oil and 1 egg. Mix well.
3. Remove any labels or decorations from can. Grease can well with some shortening or margarine on a paper towel. (Don't spray can with cooking spray - recipe will not work in a sprayed can.)
4. Spoon batter into can. Bake 28 to 32 minutes at 350°. Accurate oven temp is important. Cake is done when a broom straw or bamboo skewer placed in middle of cake comes out with no wet batter sticking to it.
5. While cake is baking, empty frosting mix into a small bowl. Add 2 Tb. soft butter or margarine and 1 Tb. water to frosting mix. Stir well.
6. When cake is done let it cool 15 minutes in can. Run plain table knife around edge of cake to loosen it from can. Shake can to allow cake to slide out.
7. To serve, cut cake into slices the desired thickness. Top each serving with some of the frosting.

Pina Colada Cake Mix In A Can

1 pineapple cake mix (Betty Crocker, Duncan Hines, etc.)
1 (4-serving size) coconut cream instant pudding mix

1. Place ingredients into large bowl. Blend with whisk.
2. Measure 3/4 cup of this mixture into a zipper sandwich bag. Repeat. (You should get 6 bags of mix from this cake mix/ pudding mix blend.)

Coconut Frosting Mix

(Make 1 Frosting Mix for each Pina Colada Cake Mix.)

3/4 cup powdered sugar
1/3 cup flaked coconut

3. Place frosting mix ingredients into a small bowl. Blend well. Place in a zipper sandwich bag. Label this bag "Coconut Frosting Mix".
4. Place a bag of cake mix and a bag of frosting mix into each of six 14 to 16 oz. cans (see pgs. 4-5 for detailed instructions on kinds of cans and decorating cans.)
5. Attach a recipe card with the following instructions:

Pina Colada Cake Baked In A Can
(With Coconut Frosting)

1. Remove mix from can. (Set aside frosting mix packet for later use.) Empty bag of mix into a medium bowl.
2. Add 1 Tb. water, 1 Tb. oil and 1 egg. Mix well.
3. Remove any labels or decorations from can. Grease can well with some shortening or margarine on a paper towel. (Don't spray can with cooking spray - recipe will not work in a sprayed can.)
4. Spoon batter into can. Bake 28 to 32 minutes at 350°. Accurate oven temp is important. Cake is done when a broom straw or bamboo skewer placed in middle of cake comes out with no wet batter sticking to it.
5. While cake is baking, empty frosting mix into a small bowl. Add 2 Tb. soft butter or margarine and 1 Tb. water to frosting mix. Stir well.
6. When cake is done let it cool 15 minutes in can. Run plain table knife around edge of cake to loosen it from can. Shake can to allow cake to slide out.
7. To serve, cut cake into slices the desired thickness. Top each serving with some of the frosting.

Date Nut Bread Mix In A Can

1/2 cup flour
2 Tb. sugar
2 Tb. finely chopped pecans
2 Tb. chopped dates (buy a box of chopped dates)
1 Tb. brown sugar
1 tsp. baking powder

1. Place ingredients into medium bowl. Blend with whisk.
2. Place mixture into a zipper sandwich bag. Place bag into a 14 to 16 oz. can (see pgs. 4-5 for detailed instructions on kinds of cans and decorating cans.)
3. Attach a recipe card with the following instructions:

Date Nut Bread To Bake In A Can

1. Remove mix from can. Empty bag of mix into a medium bowl.
2. Add 1 Tb. water, 1 Tb. oil and 1 egg. Mix well.
3. Remove any labels or decorations from can. Grease can well with some shortening or margarine on a paper towel. (Don't spray can with cooking spray - recipe will not work in a sprayed can.)
4. Spoon batter into can. Bake 32 to 35 minutes at 350°. Accurate oven temp is important. Bread is done when a broom straw or bamboo skewer placed in middle of bread comes out with no wet batter sticking to it.
5. When bread is done let it cool 15 minutes in can. Run plain table knife around edge of bread to loosen it from can. Shake can to allow bread to slide out.
6. To serve, cut bread into slices the desired thickness.

Spice Cake Mix In A Can

1 spice cake mix (Betty Crocker Duncan Hines, etc.)

1 (4-serving size) butterscotch instant pudding mix

1. Place ingredients into large bowl. Blend with whisk.
2. Measure 3/4 cup of this mixture into a zipper sandwich bag. Repeat. (You should get 6 bags of mix from this cake mix/ pudding mix blend.)

Brown Sugar Frosting Mix

(Make 1 Frosting Mix for each Spice Cake Mix.)

1/2 cup powdered sugar

1/3 cup brown sugar

3. Place frosting mix ingredients into a small bowl. Blend well. Place in a zipper sandwich bag. Label this bag "Brown Sugar Frosting Mix".
4. Place a bag of cake mix and a bag of frosting mix into each of six 14 to 16 oz. cans (see pgs. 4-5 for detailed instructions on kinds of cans and decorating cans.)
5. Attach a recipe card with the following instructions:

Spice Cake To Bake In A Can
(With Brown Sugar Frosting)

1. Remove mix from can. (Set aside frosting mix packet for later use.) Empty bag of mix into a medium bowl.
2. Add 1 Tb. water, 1 Tb. oil and 1 egg. Mix well.
3. Remove any labels or decorations from can. Grease can well with some shortening or margarine on a paper towel. (Don't spray can with cooking spray - recipe will not work in a sprayed can.)
4. Spoon batter into can. Bake 28 to 32 minutes at 350°. Accurate oven temp is important. Cake is done when a broom straw or bamboo skewer placed in middle of cake comes out with no wet batter sticking to it.
5. While cake is baking, empty frosting mix into a small bowl. Add 1/2 cup melted butter or margarine to frosting mix. Stir until creamy.
6. When cake is done let it cool 15 minutes in can. Run plain table knife around edge of cake to loosen it from can. Shake can to allow cake to slide out.
7. To serve, cut cake into slices the desired thickness. Top each serving with some of the frosting.

Messy Milky Way Cake Mix In A Can
or
Sloppy Snickers Bar Cake Mix In A Can

3/4 cup German chocolate cake mix (Betty Crocker, Duncan Duncan Hines, etc.)

2 FUN SIZE Milky Way or Snickers Candies**

** Be sure to get "fun size", NOT "miniatures". These are small, but not as small as "miniatures". The size you want measures about 1 1/2" x 1".

1. Place cake mix into a zipper sandwich bag.
2. Place this bag into a 14 to 16 oz. can (see pgs. 4-5 for detailed instructions on kinds of cans and decorating cans.)
3. Place the wrapped candy bars into the can or tie them onto the outside of the can as part of the decorations,
4. Attach a recipe card with the following instructions: (Name recipe after whichever candy bar you are using.)

Messy Milky Way Cake To Bake In A Can

1. Remove mix from can. Empty bag of mix into a medium bowl.
2. Add 1 Tb. water, 1 Tb. oil and 1 egg. Mix well.
3. Unwrap candy bars and chop them into 4 equal size pieces each. Add candy bars to batter. Stir well.
4. Remove any labels or decorations from can. Grease can well with some shortening or margarine on a paper towel. (Don't spray can with cooking spray - recipe will not work in a sprayed can.)
5. Spoon batter into can. Bake 28 to 32 minutes at 350°. Accurate oven temp is important. Cake is done when a broom straw or bamboo skewer placed in middle of cake comes out with no wet batter sticking to it.
6. When cake is done let it cool in can for 5 minutes. Run plain table knife around edge of cake to loosen it from can. Shake can to allow cake to slide out onto a plate. The candy will be melted in the bottom of the can.

   Cut the cake into 4 long spears with some of the candy sticking to top of each spear. Spoon remaining candy out of the can onto each piece as "frosting".

   Serve warm or at room temperature.

Disappearing Marshmallow Cake Mix In A Can

3/4 cup German chocolate cake mix, (Betty Crocker, Duncan Hines,)
1/3 cup mini marshmallows

1. Place ingredients into large bowl. Blend with whisk.
2. Place mixture into a zipper sandwich bag.

Chocolate Marshmallow Glaze Mix
(Make 1 Glaze Mix for each Disappearing Cake Mix.)

1/2 cup mini marshmallows
1/3 cup powdered sugar
1/4 cup milk chocolate chips

3. Place glaze mix ingredients into a small bowl. Blend well. Place in a zipper sandwich bag. Label this bag "Glaze Mix".
4. Place a bag of cake mix and a bag of glaze mix into a 14 to 16 oz. can (see pgs. 4-5 for detailed instructions on kinds of cans and decorating cans.)
5. Attach a recipe card with the following instructions:

Disappearing Marshmallow Cake To Bake In A Can (With Chocolate Marshmallow Glaze)

1. Remove mix from can. (Set aside glaze mix packet for later use.) Empty bag of mix into a medium bowl.
2. Add 1 Tb. water, 1 Tb. oil and 1 egg. Mix well.
3. Remove any labels or decorations from can. Grease can well with some shortening or margarine on a paper towel. (Don't spray can with cooking spray - recipe will not work in a sprayed can.)
4. Spoon batter into can. Bake 28 to 32 minutes at 350°. Accurate oven temp is important. Cake is done when a broom straw or bamboo skewer placed in middle of cake comes out with no wet batter sticking to it.
5. While cake is baking, empty glaze mix into a small bowl. Add 3 Tb. soft butter or margarine to glaze mix. Microwave 20 seconds. Stir very well. Microwave 10 more seconds. Stir very well.
6. When cake is done let it cool 15 minutes in can. Run plain table knife around edge of cake to loosen it from can. Shake can to allow cake to slide out.
7. To serve, cut cake into slices the desired thickness. Top each serving with some of the glaze.

Fudge Cake Mix In A Can

1 (19.8 oz.) box Betty Crocker Fudge Brownie Mix
OR 1 (21 oz.) box Duncan Hines Chewy Fudge Family Style Brownie Mix

(Either of these brownie mixes makes 4 mixes for cans.)

1. Place 1 cup dry mix in each of 4 sandwich bags. Divide remaining mix evenly among the 4 bags. (Each bag will end up with a little more than 1 cup in it.)

Chocolate Frosting Mix

(Make 1 Frosting Mix for each Fudge Cake Mix.)

1/2 cup milk chocolate chips
1/3 cup powdered sugar

2. Mix frosting mix ingredients and place in a zipper sandwich bag.
3. Label this bag "Frosting Mix".
4. Place a bag of cake mix and a bag of frosting mix into a 14 to 16 oz. can (see pgs. 4-5 for info on kinds of cans and decorating cans.)
5. Attach a recipe card with the following instructions:

Fudge Cake Baked In A Can
(With Chocolate Frosting)

1. Remove mix from can. (Set aside frosting mix packet for later use.) Empty bag of mix into medium bowl.
2. Add 2 Tb. oil and 1 egg. Mix well.
3. Remove any labels or decorations from can. Grease can well with some shortening or margarine on a paper towel. (Don't spray can with cooking spray - recipe will not work in a sprayed can.)
4. Spoon batter into can. Bake 55 to 65 minutes at 300°. Accurate oven temp is important. Cake is done when a broom straw or bamboo skewer placed in middle of cake comes out with no wet batter sticking to it.
5. While cake is baking, empty frosting mix into a small bowl. Add 1 Tb. water to frosting mix. Stir well. Add 3 Tb. soft butter or margarine. Microwave 20 seconds. Stir very well. Microwave 20 more seconds. Stir very well.
6. When cake is done let it cool 15 minutes in can. Run plain table knife around edge of cake to loosen it from can. Shake can to allow cake to slide out.
7. To serve, cut cake into slices the desired thickness. Top each serving with some of the frosting.

Chocolate Italian Cream Cake Mix In A Can

1 German chocolate cake mix, Betty (Crocker, Pillsbury, etc.)
1 cup flaked coconut
1 cup chopped pecans

1. Place ingredients into large bowl. Blend with whisk.
2. Measure 3/4 cup of this mixture into a zipper sandwich bag. Repeat. (You should get 6 bags of mix from this cake mix/coconut/pecan blend.)

Chocolate Coconut Frosting Mix

(Make 1 Frosting Mix for each Chocolate Italian Cream Cake Mix.)

3/4 cup powdered sugar
1/3 cup flaked coconut
2 tsp. cocoa powder
1 Tb. finely chopped pecans

3. Place frosting mix ingredients into a small bowl. Blend well. Place in a zipper sandwich bag. Label this bag "Frosting Mix".
4. Place a bag of cake mix and a bag of frosting mix into each of six 14 to 16 oz. cans (see pgs. 4-5 for detailed instructions on kinds of cans and decorating cans.)
5. Attach a recipe card with the following instructions:

Chocolate Italian Cream Cake To Bake In A Can
(With Chocolate Coconut Frosting)

1. Remove mix from can. (Set aside frosting mix packet for later use.) Empty bag of mix into a medium bowl.
2. Add 1 Tb. water, 1 Tb. oil and 1 egg. Mix well.
3. Remove any labels or decorations from can. Grease can well with some shortening or margarine on a paper towel. (Don't spray can with cooking spray - recipe will not work in a sprayed can.)
4. Spoon batter into can. Bake 28 to 32 minutes at 350°. Accurate oven temp is important. Cake is done when a broom straw or bamboo skewer placed in middle of cake comes out with no wet batter sticking to it.
5. While cake is baking, empty frosting mix into a small bowl. Add 2 Tb. soft butter or margarine and 1 Tb. water to frosting mix. Stir well.
6. When cake is done let it cool 15 minutes in can. Run plain table knife around edge of cake to loosen it from can. Shake can to allow cake to slide out.
7. To serve, cut cake into slices the desired thickness. Top each serving with some of the frosting.

Mexican Chocolate Cake Mix In A Can

3/4 cup German chocolate cake mix, (Betty Crocker, Pillsbury, etc.)
2 tsp. cinnamon

1. Mix ingredients and place in zipper sandwich bag.

Mexican Chocolate Glaze Mix

(Make 1 Glaze Mix for each Mexican Chocolate Cake Mix.)

3/4 cup powdered sugar
2 tsp. cocoa powder
1/4 tsp. cinnamon

2. Place glaze mix ingredients into a small bowl. Blend well. Place in a zipper sandwich bag. Label this bag "Glaze Mix".
3. Place a bag of cake mix and a bag of glaze mix into a 14 to 16 oz. can (see pgs. 4-5 for detailed instructions on kinds of cans and decorating cans.)
4. Attach a recipe card with the following instructions:

Mexican Chocolate Cake To Bake In A Can

(With Mexican Chocolate Glaze)

1. Remove mix from can. (Set aside glaze mix packet for later use.) Empty bag of mix into a medium bowl.
2. Add 1 Tb. water, 1 Tb. oil and 1 egg. Mix well.
3. Remove any labels or decorations from can. Grease can well with some shortening or margarine on a paper towel. (Don't spray can with cooking spray - recipe will not work in a sprayed can.)
4. Spoon batter into can. Bake 28 to 32 minutes at 350°. Accurate oven temp is important. Cake is done when a broom straw or bamboo skewer placed in middle of cake comes out with no wet batter sticking to it.
5. While cake is baking, empty glaze mix into a small bowl. Add 1 to 1 1/2 Tb. water to glaze mix. Stir well.
6. When cake is done let it cool 15 minutes in can. Run plain table knife around edge of cake to loosen it from can. Shake can to allow cake to slide out.
7. To serve, cut cake into slices the desired thickness. Top each serving with some of the glaze.

Mocha Cake Mix In A Can

1 German chocolate cake mix (Betty Crocker, Duncan Hines, etc.)
3 Tb. instant coffee granules
1 (4-serving size) chocolate instant pudding mix

1. Place ingredients into large bowl. Blend with whisk.
2. Measure 3/4 cup of this mixture into a zipper sandwich bag. Repeat. (You should get 6 bags of mix from this cake mix/ pudding mix blend.)

Mocha Frosting Mix

(Make 1 Frosting Mix for each Mocha Cake Mix.)

3/4 cup powdered sugar
2 Tb. cocoa powder
1/2 tsp. instant coffee granules

3. Place frosting mix ingredients into a small bowl. Blend well. Place in zipper sandwich bag. Label this "Mocha Frosting Mix".
4. Place a bag of cake mix and a bag of frosting mix into each of six 14 to 16 oz. cans (see pgs. 4-5 for detailed instructions on kinds of cans and decorating cans.)
5. Attach a recipe card with the following instructions:

Mocha Cake To Bake In A Can
(With Mocha Frosting)

1. Remove mix from can. (Set aside frosting mix packet for later use.) Empty bag of mix into a medium bowl.
2. Add 1 Tb. water, 1 Tb. oil and 1 egg. Mix well.
3. Remove any labels or decorations from can. Grease can well with some shortening or margarine on a paper towel. (Don't spray can with cooking spray - recipe will not work in a sprayed can.)
4. Spoon batter into can. Bake 28 to 32 minutes at 350°. Accurate oven temp is important. Cake is done when a broom straw or bamboo skewer placed in middle of cake comes out with no wet batter sticking to it.
5. While cake is baking, empty frosting mix into a small bowl. Add 2 Tb. soft butter and 1 to 2 Tb. water to frosting mix. Stir until smooth and creamy.
6. When cake is done let it cool 15 minutes in can. Run plain table knife around edge of cake to loosen it from can. Shake can to allow cake to slide out.
7. To serve, cut cake into slices the desired thickness. Top each serving with some of the frosting.

Chocolate/Dried Cranberry Cake Mix In A Can

3/4 cup German chocolate cake mix, (Betty Crocker, Pillsbury, etc.)
2 Tb. dried cranberries
2 Tb. finely chopped nuts

1. Mix ingredients and place in zipper sandwich bag.

Chocolate Frosting Mix

(Make 1 Frosting Mix for each Chocolate/Cranberry Cake Mix.)

1/2 cup milk chocolate chips
1/3 cup powdered sugar

2. Mix frosting mix ingredients and place in a zipper sandwich bag.
3. Label this bag "Frosting Mix".
4. Place a bag of cake mix and a bag of frosting mix into a 14 to 16 oz. can (see pgs. 4-5 for info on kinds of cans and decorating cans.)
5. Attach a recipe card with the following instructions:

Chocolate/Dried Cranberry Cake To Bake In A Can (With Chocolate Frosting)

1. Remove mix from can. (Set aside frosting mix packet for later use.) Empty bag of mix into a medium bowl.
2. Add 1 Tb. water, 1 Tb. oil and 1 egg. Mix well.
3. Remove any labels or decorations from can. Grease can well with some shortening or margarine on a paper towel. (Don't spray can with cooking spray - recipe will not work in a sprayed can.)
4. Spoon batter into can. Bake 28 to 32 minutes at 350°. Accurate oven temp is important. Cake is done when a broom straw or bamboo skewer placed in middle of cake comes out with no wet batter sticking to it.
5. While cake is baking, empty frosting mix into a small bowl. Add 1 Tb. water to frosting mix. Stir well. Add 3 Tb. soft butter or margarine. Microwave 20 seconds. Stir very well. Microwave 20 more seconds. Stir very well.
6. When cake is done let it cool 15 minutes in can. Run plain table knife around edge of cake to loosen it from can. Shake can to allow cake to slide out.
7. To serve, cut cake into slices the desired thickness. Top each serving with some of the frosting.

Onion Dill Bread Mix In A Can

2/3 cup Bisquick
2 tsp. Lipton Recipe Secrets
Golden Onion Soup Mix
1/4 tsp. dried dillweed
1/8 tsp. coarse black pepper

1. Place ingredients into medium bowl. Blend with whisk.
2. Place mixture into a zipper sandwich bag. Place bag into a 14 to 16 oz. can (see pgs. 4-5 for detailed instructions on kinds of cans and decorating cans.)
3. Place an additional tsp. dried dillweed into a zipper bag. Label this "Dill For Butter". Place in can.
4. Attach a recipe card with the following instructions:

Onion Dill Bread To Bake In A Can (With Dill Butter)

1. Remove mix from can. (Set aside dill packet for later use.) Empty bag of mix into a medium bowl.
2. Add 3 Tb. milk and 1 egg white. Mix well.
3. Remove any labels or decorations from can. Grease can well with some shortening or margarine on a paper towel. (Don't spray can with cooking spray - recipe will not work in a sprayed can.) Place empty can in preheated oven for 5 minutes to heat can.
4. Spoon batter into can. Bake 23 to 26 minutes at 375°. Accurate oven temp is important. Bread is done when a broom straw or bamboo skewer placed in middle of bread comes out with no wet batter sticking to it.
5. While bread is baking, mix dill from dill packet into 1/2 stick soft butter or margarine.
6. When bread is done let it cool 15 minutes in can. Run plain table knife around edge of cake to loosen it from can. Shake can to allow bread to slide out.
7. To serve, cut bread into slices the desired thickness and serve with dill butter.

Cornbread Mix In A Can

1/3 cup yellow cornmeal
1/4 cup flour
1 Tb. sugar
1 tsp. baking powder
1/4 tsp. salt

1. Place ingredients into medium bowl. Blend with whisk.
2. Place mixture into a zipper sandwich bag. Place bag into a 14 to 16 oz. can (see pgs. 4-5 for detailed instructions on kinds of cans and decorating cans.)
3. Attach a recipe card with the following instructions:

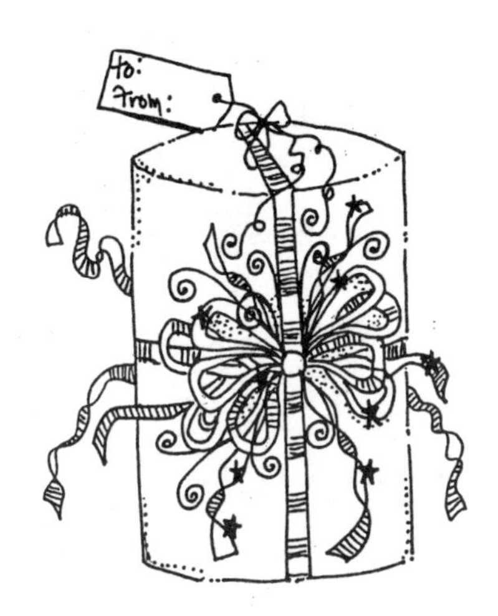

Cornbread To Bake In A Can

1. Remove mix from can. Empty bag of mix into a medium bowl.
2. Add 1/3 cup milk, 1 Tb. oil and 1 egg white. Mix well.
3. Remove any labels or decorations from can. Grease can well with some shortening or margarine on a paper towel. (Don't spray can with cooking spray - recipe will not work in a sprayed can.) Place empty can in preheated oven for 5 minutes to heat can.
4. Spoon batter into can. Bake 32 to 35 minutes at 450°. Accurate oven temp is important. Bread is done when a broom straw or bamboo skewer placed in middle of bread comes out with no wet batter sticking to it.
5. When bread is done let it cool in can for 15 minutes. Run plain table knife around edge of bread to loosen it from can. Shake can to allow bread to slide out.
6. To serve, cut bread into slices the desired thickness.

Parmesan Herb Cornbread Mix In A Can

1/3 cup yellow cornmeal
1/4 cup flour
2 Tb. parmesan (green can)
1 Tb. sugar
1 tsp. baking powder
1/4 tsp. salt
1/2 tsp. herb blend (like Italian herbs or Mrs. Dash - any herb blend with no salt

1. Place ingredients into medium bowl. Blend with whisk.
2. Place mixture into a zipper sandwich bag. Place bag into a 14 to 16 oz. can (see pgs. 4-5 for detailed instructions on kinds of cans and decorating cans.)
3. Place 1 tsp. additional herbs into a separate zipper bag. Label this "Herbs for Butter".
4. Attach a recipe card with the following instructions:

Parmesan Herb Cornbread To Bake In A Can (With Herb Butter)

1. Remove mix from can. Set aside herb packet for later use. Empty bag of mix into a medium bowl.
2. Add 1/3 cup milk, 1 Tb. oil and 1 egg white. Mix well.
3. Remove any labels or decorations from can. Grease can well with some shortening or margarine on a paper towel. (Don't spray can with cooking spray - recipe will not work in a sprayed can.) Place empty can in preheated oven for 5 minutes to heat can.
4. Spoon batter into can. Bake 23 to 26 minutes at 450°. Accurate oven temp is important. Bread is done when a broom straw or bamboo skewer placed in middle of bread comes out with no wet batter sticking to it.
5. While bread is baking, mix herbs from herb packet into 1/2 stick soft butter or margarine.
6. When bread is done let it cool 15 minutes in can. Run plain table knife around edge of bread to loosen it from can. Shake can to allow bread to slide out.
7. To serve, cut bread into slices the desired thickness and serve with herb butter.

Oatmeal Bread Mix In A Can

2/3 cup Bisquick
1/3 cup quick oats
1/4 cup brown sugar
1/2 tsp. cinnamon

1. Place ingredients into medium bowl. Blend with whisk.
2. Place mixture into a zipper sandwich bag. Place bag into a 14 to 16 oz. can (see pgs. 4-5 for detailed instructions on kinds of cans and decorating cans.)
3. Attach a recipe card with the following instructions:

Oatmeal Bread To Bake In A Can

1. Remove mix from can. Empty bag of mix into a medium bowl.
2. Add 2 Tb. water, 1 Tb. oil and 1 egg. Mix well.
3. Remove any labels or decorations from can. Grease can well with some shortening or margarine on a paper towel. (Don't spray can with cooking spray-recipe will not work in a sprayed can.)
4. Spoon batter into can. Bake 33 to 36 minutes at 375°. Accurate oven temp is important. Bread is done when a broom straw or bamboo skewer placed in middle of bread comes out with no wet batter sticking to it.
5. When bread is done let it cool in can 15 minutes. Run plain table knife around edge of bread to loosen it from can. Shake can to allow bread to slide out.
6. To serve, cut bread into slices the desired thickness.

Granola Bread Mix In A Can

2/3 cup Bisquick
1/2 cup Post Raisin Bran cereal
2 Tb. finely chopped nuts
2 Tb. flaked coconut
1 Tb. sugar
1 Tb. brown sugar
1/4 tsp. cinnamon

1. Place ingredients into medium bowl. Blend well.
2. Place mixture into a zipper sandwich bag. Place bag into a 14 to 16 oz. can (see pgs. 4-5 for detailed instructions on kinds of cans and decorating cans.)
3. Attach a recipe card with the following instructions:

Granola Bread To Bake In A Can

1. Remove mix from can. Empty bag of mix into a medium bowl.
2. Add 1 Tb. water, 1 Tb. oil and 1 egg. Mix well.
3. Remove any labels or decorations from can. Grease can well with some shortening or margarine on a paper towel. (Don't spray can with cooking spray - recipe will not work in a sprayed can.)
4. Spoon batter into can. Bake 33 to 37 minutes at 375°. Accurate oven temp is important. Bread is done when a broom straw or bamboo skewer placed in middle of bread comes out with no wet batter sticking to it.
5. When bread is done let it cool 15 minutes in can. Run plain table knife around edge of bread to loosen it from can. Shake can to allow bread to slide out.
6. To serve, cut bread into slices the desired thickness.

Beer Bread Mix In A Can

3/4 cup Bisquick
1 Tb. sugar
1 tsp. Italian herb seasoning (opt.)

1. Place ingredients into medium bowl. Blend with whisk.
2. Place into a zipper sandwich bag. Place bag into a 14 to 16 oz. can (see pgs. 4-5 for detailed instructions on kinds of cans and decorating cans.)
3. Attach a recipe card with the following instructions:

Beer Bread To Bake In A Can

1. Remove mix from can. Empty bag of mix into a medium bowl.
2. Add 1/3 cup beer. Mix well for 30 seconds,.
3. Remove any labels or decorations from can. Grease can well with some shortening or margarine on a paper towel. (Don't spray can with cooking spray-recipe will not work in a sprayed can.) Place empty can in preheated oven for 5 minutes to heat can.
4. Spoon batter into can. Bake 32 to 35 minutes at 375°. Accurate oven temp is important. Bread is done when a broom straw or bamboo skewer placed in middle of bread comes out with no wet batter sticking to it.
5. When bread is done let it cool 15 minutes in can. Run plain table knife around edge of bread to loosen it from can. Shake can to allow bread to slide out.
6. To serve, cut bread into slices the desired thickness.

White Dinner Bread Mix In A Can

1/3 cup Bisquick
1/3 cup white cornmeal
1 Tb. sugar
1/2 tsp. baking powder
1/8 tsp. salt

1. Place ingredients into medium bowl. Blend with whisk.
2. Place mixture into a zipper sandwich bag. Place bag into a 14 to 16 oz. can (see pgs. 4-5 for detailed instructions on kinds of cans and decorating cans.)
3. Attach a recipe card with the following instructions:

White Dinner Bread To Bake In A Can

1. Remove mix from can. Empty bag of mix into a medium bowl.
2. Add 1/3 cup milk, 1 Tb. oil and 1 egg white. Mix well.
3. Remove any labels or decorations from can. Grease can well with some shortening or margarine on a paper towel. (Don't spray can with cooking spray - recipe will not work in a sprayed can.)
4. Spoon batter into can. Bake 23 to 26 minutes at 375°. Accurate oven temp is important. Cake is done when a broom straw or bamboo skewer placed in middle of cake comes out with no wet batter sticking to it.
5. When bread is done let it cool 15 minutes in can. Run plain table knife around edge of cake to loosen it from can. Shake can to allow bread to slide out.
6. To serve, cut bread into slices the desired thickness and serve with butter.

Jam Swirl Bread Mix In A Can

2/3 cup Bisquick
1/2 tsp. cinnamon

1. Place ingredients into medium bowl. Blend with whisk.
2. Place mixture into a zipper sandwich bag. Place bag into a 14 to 16 oz. can (see pgs. 4-5 for detailed instructions on kinds of cans and decorating cans.)
3. Attach a recipe card with the following instructions:

NOTE: You can do this mix two ways - you can ask the recipient to add her own jam (that is the way their instructions are written now) OR you can place 1/4 cup jam into a very small baby food jar. Place jar into can. Reword instruction #2 in their instructions to say "add jam from jar and only barely. . . . . "

Jam Swirl Bread To Bake In A Can

1. Remove mix from can. Empty mix bag into a medium bowl.
2. Add 3 Tb. milk, 1 egg white. Mix well. Add 1/4 cup jam or preserves (any flavor) and only barely mix - you don't want it blended - just swirled.
3. Remove any labels or decorations from can. Grease can well with some shortening or margarine on a paper towel. (Don't spray can with cooking spray-recipe will not work in a sprayed can.) Place empty can in preheated oven to heat can. Heat can for 5 minutes.
4. Spoon batter into can. Bake 28 to 32 minutes at 375°. Accurate oven temp is important. Bread is done when a broom straw or bamboo skewer placed in middle of bread comes out with no wet batter sticking to it.
5. When bread is done let it cool 15 minutes in can. Run plain table knife around edge of bread to loosen it from can. Shake can to allow bread to slide out.
6. To serve, cut bread into slices the desired thickness.

Cinnamon Swirl Bread Mix In A Can

2/3 cup Bisquick
1 Tb. sugar

1 Tb. brown sugar
1 Tb. finely chopped pecans
1/2 tsp. cinnamon

1. Mix Bisquick and sugar and place in zipper sandwich bag.
2. Mix next three ingredients. Place this mixture into a separate zipper bag. Label this "Streusel Mix".
3. Place a bag of bread mix and a bag of streusel mix into a 14 to 16 oz. can (see pgs. 4-5 for detailed instructions on kinds of cans and decorating cans.)
4. Attach a recipe card with the following instructions:

Cinnamon Swirl Bread To Bake In A Can

1. Remove mix from can. Empty bag of mix into a medium bowl.
2. Add 3 Tb. milk and 1 egg white. Mix well.
3. Sprinkle streusel mix from streusel mix packet into batter. Stir once or twice to just barely mix.
4. Remove any labels or decorations from can. Grease can well with some shortening or margarine on a paper towel. (Don't spray can with cooking spray-recipe will not work in a sprayed can.)
5. Spoon batter into can. Bake 28 to 32 minutes at 375°. Accurate oven temp is important. Bread is done when a broom straw or bamboo skewer placed in middle of bread comes out with no wet batter sticking to it.
6. When bread is done let it cool 15 minutes in can. Run plain table knife around edge of bread to loosen it from can. Shake can to allow bread to slide out.
7. To serve, cut bread into slices the desired thickness.

Decorate A Can To Look Like A *Doll* Wearing A Lace Skirt.

You really need a computer with some clip art programs to find a picture of a doll or a picture of a girl to be the doll. For the doll photo on pg. 7 we used PrintMaster, published by Broderbund.

For her dress buy some 4"to 5" wide pre-gathered lace or eyelet. Put a paper label around the can in a color that works with your color scheme. Hot glue the lace around the top of the can. (Glue onto the can where the gathered stitches of the lace will go. That way the narrow edge of the gathered lace will be at the top of the can and the long part of the lace will form her "skirt" around the can.) Glue a ribbon with a bow around the gathered stitches - this will represent her sash.

Put the mix into the can. Top it with some Easter "grass" in a coordinating color. For the doll you will need a print-out on card weight paper of the upper body and head of a cartoon like girl (or you can use a actual photo of a girl standing up straight and looking straight ahead). Her upper body and head needs to be printed in color about 6" tall so she will be in proportion to her lace "skirt". Tape this picture of the doll inside the back inside edge of the can so she stands up at the back of the can and the lace around the can looks like her skirt.

Index